Fraction Ideas And Possibilities

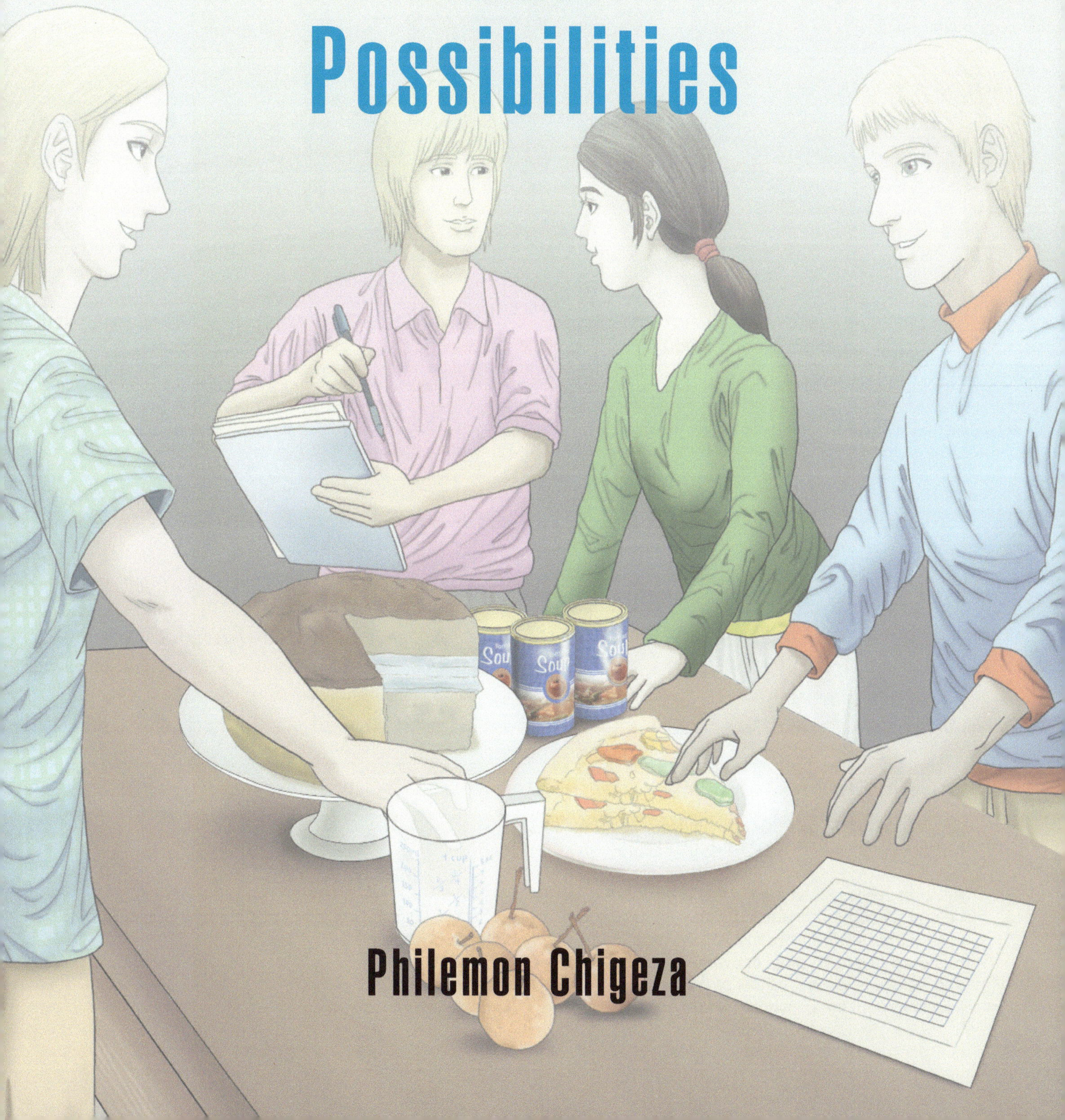

Philemon Chigeza

To order additional copies of this book, contact:
Xlibris Corporation
1-800-618-969
www.xlibris.com.au
Orders@Xlibris.com.au

Contents

CHAPTER 1
Fraction Ideas

"I have never been good at fractions," confesses Ron to his colleagues Tim, Liz and Pen. The four middle school students are working on their project on fraction ideas, challenges and possibilities due at the end of the semester.

"What do you mean when you say you have never been good at fractions Ron?" Tim inquires. "Do you mean you have never taken the time to understand fraction ideas because we use them every day?"

"What are some of the fraction ideas we use every day?" asks Ron.

"Fractions are numbers representing parts of things. The fraction ideas we use every day involves different ways we divide and share equal parts of things," responds Pen. "And how we compare the parts, whether they are equal or not."

"The language we use when describing fractions is also very important," Liz adds quickly. "For example, the idea of partitioning, this is cutting the whole into equal parts."

Tim looks in his notebook and points out, "The language we use to describe and name fraction depends on the number of equal parts that are cut from the whole. If you cut into 5 equal parts you get fifths, and if you cut into 6 equal parts you get sixths."

"You say we cut or divide the whole. What is this whole we cut or divide," asks Ron.

"This can be a whole shape as in a paper rectangle, square, circle, hexagon, or any other regular plane shape. It can also be food items like a chocolate bar, cake, pizza, banana, orange or any other fruit," suggests Pen. "We represent a fraction as the partitioning of the whole object, whatever the object is. It is important to see a whole in all its representational forms."

"This idea of partitioning or dividing into equal parts is very important," observes Tim. "If you do not cut the pizza or orange into eight equal parts, then we talk of sectors."

"Another interesting notion is that when you increase the number of equal divisions, the sizes of the cake, pizza or orange gets smaller and smaller. This is what they call the inverse relationship between the number of divisions or parts and the size of each division or part," Liz demonstrates how cutting into more parts reduces the size of the parts on a paper circle.

"The fraction ideas do not seem too difficult to understand," points out Ron. "Dividing things into equal parts, and the more parts you divide into, the smaller the parts."

Image 1: chocolate bar, cake, pizza and banana.

"Dividing the whole into parts is not the only way we represent fractions," advises Tim. "Fractions can also be represented as a collection of things that are not meant to be divided further."

Pen elaborates, "This can be like a collection of cards, a bag of lollies or marbles, a dozen eggs, or even our class which has 25 students, 15 girls and 10 boys. We can represent the number of girls as a fraction of the class, it is 15 out of 25, or the number of boys as a fraction of the class is 10 out of 25."

"This representation of fractions as a collection of things is very useful because we can apply it to many contexts," observes Liz.

Image 2: a collection of cards, a bag of lollies, marbles and a dozen eggs.

"Writing or recording common fractions is another important thing. For example one part out of four (1 out of 4), which is also called one quarter, can be recorded as (¼)," Pen points out. "That is one number over another with a line in between. The number at the bottom is called the denominator and tells how many parts the whole is divided into. The number at the top is called the numerator and tells how many parts are taken."

"The mathematical meaning of ¼ is 1 divided by 4 which is 0.25 (i.e. 1 ÷ 4 = 0.25). This representation of a common fraction as a number is called a decimal fraction," explains Tim. "It is important to understand the links between common fractions, decimal fractions and other types of fractions and representations. Understanding fraction ideas help us to learn advanced mathematical ideas."

CHAPTER 2
Representing Fractions

"So what challenges can we meet learning about fractions," inquires Ron.

"The challenges we meet can include reading, naming, ordering, representing and applying different types of fraction," responds Tim.

"It is important to understand the many representations and meanings of fractions, their connectedness and applications," argues Tim. "This includes the different ways of expressing and representing common fractions, decimal fractions, per cent, improper fractions and mixed numbers. No one representation is sufficient on its own."

"We have already started exploring common fractions and decimal fractions. How do we distinguish common fractions from other types of fractions," Ron seeks clarification from his colleagues.

"We can compare the size of the numerator and denominator; one quarter (¼) is a common fraction because the numerator is less than the denominator. If a fraction has a numerator that is less than the denominator, it is a common fraction," Liz clarifies.

"We have explored the links between common fractions and decimal fractions, but what about per cent and why is this link important," asks Ron.

"We explained that one quarter ($\frac{1}{4}$) is 1 divided by 4 which is 0.25 and that $\frac{1}{4}$ is a common fraction and 0.25 is a decimal fraction," Tim points out. "More generally decimal fractions are fractions with denominators that are 10, 100, 1000 or higher powers of 10."

Pen quickly adds, "Expressing fractions as decimals makes it very easy to represent the fractions on the number line. Representing fractions on the number line enables us to easily compare them. If you have a number of fractions, for example $\frac{2}{5}$; $\frac{3}{8}$; $\frac{4}{9}$; $\frac{3}{10}$ and $\frac{4}{11}$, changing them into decimal fractions and representing them on the number line makes it easy to compare them, that is identify which one is the biggest and smallest."

Liz points at a calculator and suggests, "You can use a calculator to divide the numerator by the denominator to change the common fractions to decimal fractions and then represent the decimal fractions on the number line."

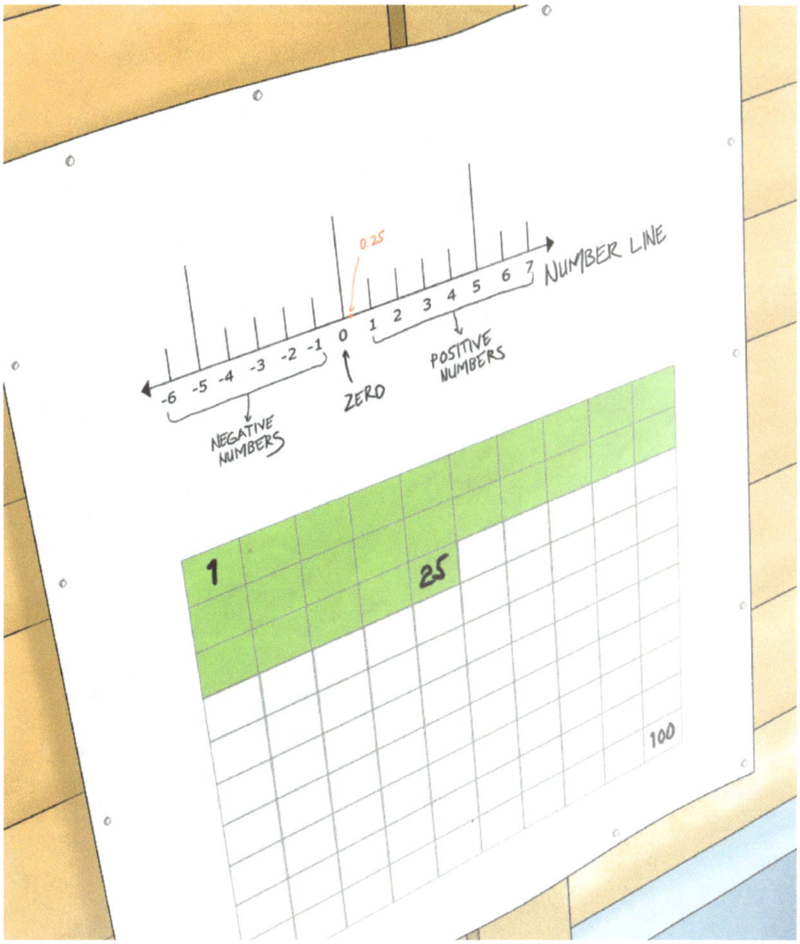

Change the common fractions $\frac{2}{5}$; $\frac{3}{8}$ and $\frac{4}{9}$ into decimal fractions and represent them on a number line.

"This idea of fractions as a measure on the number line is also very helpful when you estimate measurements on most scales we use," adds Tim.

Image 3: a number line.

Pen looks in her notebook and advices, "When the denominator of a fraction is 100, then it is a per cent. The symbol for per cent is (%) and so 25 out of 100 can be written as 25%. This can be represented by the 25 coloured small squares out of the 100 small squares in the box." She points at an illustration of the box.

"The 25% can also be represented by 25 green lollies in a jar containing 100 lollies," adds Tim.

"You also have to realise that 25% or 25 out of 100 also means 25 ÷ 100 which is equal to 0.25 which represents a decimal fraction. And the number 0.25 is read as *twenty-five hundredths*," explains Liz.

"We have agreed that it is important to show the links between common fractions and decimal fractions and percentages," points out Ron. "And represent the fractions on the number line because it helps us compare fractions and identify which one is big, small or equivalent. I still have a problem with equivalent fractions."

"When working with common fractions, equivalent fractions often need to be found. Equivalent fractions have the same value, even though they may look different," observes Tim.

"A good starting point to understand equivalent fractions is to investigate different representations of the same amount as shown in the fraction chart below. The chart shows that ½ is equivalent to 2 quarters (2 × ¼)."

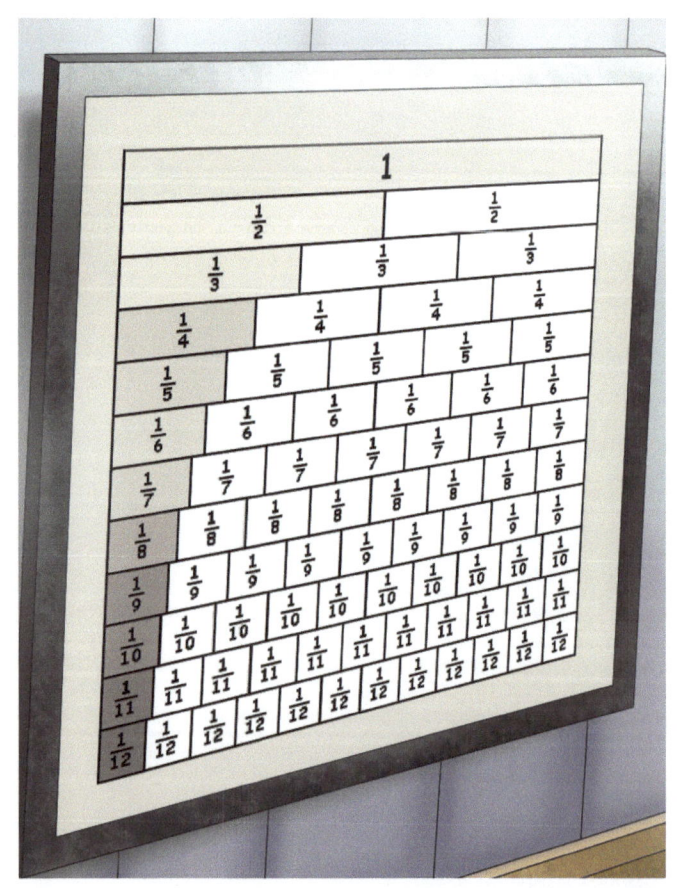

Image 4: of the equivalent fraction chart.

"What are mixed numbers and improper fractions," asks Ron.

"A mixed number is a number that has a whole number and a fraction, for example 2¾. The number 2 is the whole and ¾ is the fraction which can also be represented as a decimal fraction 0.75. This means that 2¾ can be written as 2.75 and can be represented on the number line," explains Liz.

"For improper fractions, the numerator is bigger than the denominator. The mixed number 2¾ as an improper fraction is $\frac{11}{4}$," adds Tim.

"Can we use a calculator to add, subtract, multiply or divide common fractions, improper fractions and mixed numbers," asks Ron.

"Yes we can use a calculator to quickly add, subtract, multiply or divide common fractions, improper fractions and mixed numbers though there are some other methods we can use," suggests Liz. "But we have to change the fractions into their decimal form first, that is change 2¾ to 2.75. Calculators work with decimal numbers."

CHAPTER 3
Applying Fraction Ideas

"Tim proposed that understanding fraction ideas helps us to learn and develop other mathematical ideas. What are some of these mathematical ideas," Ron asks his colleagues.

"One such example involves the idea of ratio and proportion," responds Tim. "Ratio relates to fraction ideas and sometimes is taken as another representation of fractions."

"A ratio compares quantities of the same kind in a definite order. This is why a ratio is stated without using units," explains Liz. "We usually write ratios in their simplest form using whole numbers, in the same way we write fractions in simplest form."

"Imagine you are making fruit punch for a party and the recipe requires one part fruit juice to three parts ginger ale. This is a ratio of 1 is to 3 (i.e. 1:3). You are supposed to mix the fruit punch by keeping the ratio (or component parts) of the juice and ginger ale constant," elaborates Pen.

"It is important to keep the unit of measuring capacity constant when making the fruit punch," explains Tim. "This means if you are using a measuring cup, keep the measure of the juice and ginger ale the same."

"The juice and ginger must remain in the same ratio," advices Pen.

Complete the table and draw a line graph with the amount of fruit juice on the horizontal axis and the amount of ginger ale on the vertical axis.

Fruit juice (units)	1	1.5		3.5	4
Ginger ale (units)			6		15

Image 5: fruit juice, ginger ale, measuring cup and bowl.

"Closely related to ratio is the idea of rate," suggests Liz. "A rate is obtained by dividing one quantity by a different related quantity."

Pen looks in her notebook and explains, "The measurement units for the quantities in a rate are separated by the word 'per' or the symbol /, with the second one being the denominator."

"Most rates are prices, pay rates and speeds," Tim elaborates. "A fruit and vegetable shop shows the price of tomatoes as a rate, for example the price of tomatoes in our local shop is marked as $5.40 per kg."

"This rate tells how much you pay for each kilogram. For example, if you buy 2kg of tomatoes from the shop you pay $10.80, and if you buy 3kg you pay $16.20," Liz confirms the calculations on her calculator.

Image 6: prices of fruit.

"The idea of gradient is closely related to rate," proposes Tim. "The gradient or slope of a straight line is a measure of how steep the line is and can be thought of as the ratio of the rise or fall to the run."

"When you drive on a steep road, normally there is a road sign that indicates the amount of gradient or slope on the road," adds Pen.

Image 7: steep road.

"Direct proportion is yet another mathematical idea closely related to ratio. When one quantity is directly proportional to another, then if it changes, the other changes by the same factor," points out Liz.

"For example, suppose that you are buying cans of soup at the shop and a can of soup costs $0.50. If you buy 4 cans, you would pay $2.00. If you buy 8 cans, you would pay $4.00. Therefore we can conclude that twice the soup will cost you twice the money," explains Tim.

CHAPTER 4
Per cent and Money

"In everyday life, fraction ideas most commonly used are in the form of per cents (%)," observes Pen. "Per cent is used to calculate discount, sales tax, interest on investment and other financial transactions."

"We have already explained that the term 'per cent' means 'out of hundred' and the symbol is %," Tim recalls. "Every day we see advertisement like: discount or sale 40% off, earn 10% on your money or school attendances down by 12%."

"An understanding of the notion of per cent is very important for all citizens to be able to function competently with their everyday activities," adds Ron.

"When calculating discount, sales tax, interest on investment or any other transaction, the most important step is to be aware of the amount that is taken to be 100% and represents a whole," suggests Liz. "Usual 100% refers to the initial value of that quantity."

"It is also important to be clear that a 10% increase or a 10% decrease in a quantity is usually relative to the initial value of that quantity," adds Tim.

"If an item is initially priced at $200 and the price rises by 10% which is an increase of $20, the new price will be $220." Pen elaborates. "It is important to realise that this final price is 110% of the initial price (i.e. 100% + 10% = 110%)."

"From this logic, an increase of 100% in a quantity means that the final amount is 200% of the initial amount (100% + 100% = 200%). The final amount is doubled," Liz clarifies.

Image 8: advertisement sale 40% off on items.

"Per cent is used to describe increases or decreases of prices when shopping and in other financial situations," highlights Ron.

"For example I was given a discount of 20% on this notebook and its marked price was $400.00," explains Liz. "To get the actual discount I was given, you find 20% of $400, which is $80.00."

Pen does a quick calculation and suggests, "So you paid $320.00, which is the difference between $400.00 and $80.00."

"When calculating simple interest, the initial sum of money invested called the principal represents the 100%. The amount of interest is calculated on the principal," points out Tim. "Simple interest = percentage rate x principal x time."

"To find simple interest on $1000.00 invested for 3 years at an interest rate of 5%: Simple interest is 5/100 x 1000 x 3 = 150," explains Liz. "The total sum after 3 years is $1000.00 + $150.00 which is equal to $1150.00."

Image 9: Bank advertisement.

"If the interest is not withdrawn until the end of the entire period then compound interest must be calculated," explains Pen. "Here the principal increases year on year and causes a new calculation for each year."

Tim reads from his notebook, "Find the compound interest on $1000.00 invested for 3 years at an interest rate of 5%."

"Interest for first year: = 5/100 x 1000 = 50.00. Add to principal ($1000.00 + $50.00) = $1050.00," suggests Liz after a quick calculation and encourages Pen and Tim to do the next calculations.

"Interest for second year: 5/100 x 1050 = 52.50. Add to principal ($1050.00 + $52.50) = $1102.50," adds Pen.

"Interest for third year: 5/100 x 1102.50 = 55.125, which can be rounded as $55.13. Add to principal ($1102.50 + $55.13) = $1157.63," adds Tim.

"Compound interest accumulates quicker than simple interest," observes Ron.

CHAPTER 5

Possibilities

"Fractions are numbers representing parts of things. The fraction ideas we use every day include dividing and sharing parts of things. Therefore, it becomes important to understand the links between common fractions, decimal fractions and other types of fractions and representations. Understanding fractions help us to learn advanced mathematical ideas," Pen summarises the main ideas.

"The challenges we encounter include reading, naming, ordering, representing and applying different types of fractions," explains Tim.

"It is important to understand the many representations and meanings of fractions, their connectedness and applications," suggests Liz. "Representing fractions on the number line helps us compare fractions to identify which one is big, small or equivalent."

"An understanding of fraction ideas helps us to develop the notion of ratio, which compares quantities of the same kind in a definite order, and other related concepts which include rates as in prices of food per kg or pay rates. Also gradient or slope of a straight line and direct proportion, where if one thing changes the other changes by the same factor," elaborates Pen.

Image 10: dividing the sizes of the cake, pizza and orange.

"Per cent is the most commonly used fraction idea," highlights Ron. "Per cent is used to calculate discount, sales tax, interest on investment and other financial transactions."

"It is helpful to realise that 100% refers to the initial value of the quantity," explains Liz.

"It is also helpful to realise that a 10% increase or a 10% decrease in a quantity is usually relative to the initial value of that quantity, called the principal when calculating simple interest," points out Tim.

"We all use fraction ideas every day when we divide and share parts of things and compare the parts," explains Pen.

"Therefore everyone should develop competence and proficiency when dealing with fraction ideas, as well as understand the notion of per cent which we use everyday," adds Pen. "We should explore some of these possibilities for our fraction project:

1. Explore how we can help students in our class and lower grades to develop proficiency reading, naming, ordering, representing and applying different types of fractions.

2. Explore how we can help students in our class and lower grades understand common fractions, decimal fractions and other types of fractions and representations.

3. Explore how we can help students in our class and lower grades develop competence with per cent and calculate discount, sales tax, simple and compound interest.

4. Develop a module we will use to educate the public in our community about the importance of developing competence with fraction concepts."

Get into a team with two or three friends. Choose one or more from the four possibilities of the fraction project, or any other fraction challenge. Explore how you can help students in your class, lower grades and the general public to develop competence with per cent and other fraction concepts. You are part of the generation that is going to develop fraction competence and explore possibilities for the future to make this world a better place.

www.ingramcontent.com/pod-product-compliance
Lightning Source LLC
Chambersburg PA
CBHW041311180526
45172CB00003B/1061